Garden landscape Detail Design - 3

庭院细部元素设计
观赏池、喷泉、瀑布、溪流、地形、建筑小品

③

中国林业出版社
China Forestry Publishing House

图书在版编目（CIP）数据

庭院细部元素设计.③，观赏池、喷泉、瀑布、溪流、地形、建筑小品 /《庭院细部元素设计》编委会编. —— 北京：中国林业出版社，2016.5

ISBN 978-7-5038-8502-0

Ⅰ.①庭… Ⅱ.①庭… Ⅲ.①庭院-景观设计 Ⅳ.① TU986.4

中国版本图书馆 CIP 数据核字 (2016) 第 078634 号

《庭院细部元素设计》编委会

◎ 编委会成员名单

主　　编：董　君
编写成员：董　君　李晓娟　曾　勇　梁怡婷　贾　濛　李通宇　姚美慧
　　　　　刘　丹　张　欣　钱　瑾　翟继祥　王与娟　李艳君　温国兴
　　　　　黄京娜　罗国华　夏　茜　张　敏　滕德会　周英桂　李伟进

◎ 特别鸣谢：北京吉典博图文化传播有限公司

中国林业出版社 · 建筑与家居出版中心

责任编辑：纪　亮　王思源
联系电话：010-8314 3518

出版：中国林业出版社
　　　（100009 北京西城区德内大街刘海胡同 7 号）
http://lycb.forestry.gov.cn/
E-mail：cfphz@public.bta.net.cn
电话：（010）8314 3518
发行：中国林业出版社
印刷：北京利丰雅高长城印刷有限公司
版次：2016 年 6 月第 1 版
印次：2016 年 6 月第 1 次
开本：235mm×235mm 1/12
印张：16
字数：200 千字
定价：99.00 元

鸣谢
因稿件繁多内容多样，书中部分作品无法及时联系到作者，请作者通过编辑部与主编联系获取样书，并在此表示感谢。

目 录 3
CONTENTS

观赏池　5~102

喷泉　103~132

瀑布　133~146

溪流　147~174

地形　175~258

建筑小品　259~320

GARDEN LANDSCAPE DETAIL DESIGN · VIEWING POOL
观赏池

　　庭院的意义在于居住者对它所提供的生活方式的认同，在于它所带来的院落"能指"（居住环境）和"所指"（居住者的生活方式、社会交往方式）的和谐关系，而此代码在现代集合住宅中的被扭曲和遗忘引发了居民对院落中亲密的邻里关系的怀念。

　　庭院作为人为的自然空间，它将自然和人工相结合起来，让人产生一种回归自然的向往。庭院中往往常采用小巧、精致的手法，设置植物、院路、亭廊、假山、雕塑、水池等，既表现出个性化的特点，同时给人们带来自然美、人工美的艺术享受。

Agnatus eiumendae voluptamusa custet et demporeribus minverchil mi, unte magnit eum et que doloreh enihili quaspe vendi ium nume sum vita doloria nosam as utendi te solori beribus pa consequi idias maiorem porestiis exerendias as es ut volupit, el earum labo. Itatemp orerenes ipsunt, simusap idustenet explacea et.
con rat doluptatus arum et que plit occum et iunt quam, voluptia parcius dollect aturehentum abo. Neque voluptas seque sam, qui opta volorib ernate lat latem eumet la quis qui ut arit, am reperates dolore.

Viewing pool ○ 观赏池

1.2
2.9117
3.Adelaide Zoo Giant Panda Forest
4，5.Aley, Lebanon Date of Completion-消失点
6.Austin, USA Date of Completion-串联式溪边住宅
7.Bassil Mountain Escape, Faqra, Lebanon

Viewing pool 〇 观赏池

1，2.Garden in the Silver Moonlight
3，4.Island Modern, Key West, FL
5-8.Brumadinho, Brazil Date of Completion-Burle Marx教育中心
9.Cairnlea, Victoria-Cairnlea

Viewing pool ○ 观赏池

1. Cairnlea, Victoria–Cairnlea
2, 3. Canberra, Australia Date of Completion–澳大利亚国家美术馆-澳大利亚花园
4. Canada–cSimcoe WaveDeck
5. Chicago, Illinois–The Park at Lakeshore East
6. City Park, Bradford

Viewing pool ○观赏池

1，2.Connecticut Country House, Westport, Connecticut
3.Lee Landscape, Calistoga, CA
4，5.Faqra－魔力居所FEJ
6.Holstebro, Denmark–If it is outstanding, celebrate it twice

1. Lunada Bay Residence, Palos Verdes Peninsula, Southern Californi
2. Malinalco House, Malinalco, State of Mexico, Mexico
3-7. Houston, TX-ConocoPhillips World Headquarters

Viewing pool ○ 观赏池

15

1.PA-323-010
2.Monterrey, Mexico Date of Completion-钢铁博物馆
3.Mea-Rim, Chiangmai -Baan Rimtai Saitarn
4.Padaro Lane
5.Pamet Valley
6.Plenty Road, Bundoora, Victoria-University Hill Town Centre && Wetland
7.Private ResidenceGarden of Planes, Richmond, VA

Viewing pool ○ 观赏池

Viewing pool ○ 观赏池

1，2.Speckman House Landscape, Highland Park, St. Paul, MN
3.Sonoma Vineyard, Glen Ellen, California
4，5.Taiwan, China Date of Completion-日月潭观光局管理处
6.Sonoma Vineyard, Glen Ellen, California

1,2.Vienna Way Residence, Venice, CA
3.Wittock Residence
4.Villa H. St. Gilgen
5.Washington-2200宾夕法尼亚大道
6,7.Xochimilco, Mexico City.-XOCHIMILCO ECOLOGICAL PARK

Viewing pool ○ 观赏池

Viewing pool ○ 观赏池

1，2.阿姆斯特丹-辛克尔小岛
3，4.爱尔兰-"自我"圣坛
5.爱尔兰都柏林-观海

Viewing pool ○ 观赏池

1,2.爱涛漪水园承泽苑16栋
3.爱尔兰都柏林-红杉（展示花园）
4-6.安徽省-合肥绿地国际花都二期"玫瑰苑"
7.澳大利亚悉尼-悉尼5号湿地

Viewing pool ○ 观赏池

1，2.丹麦，奥尔堡-DANISH CROWN-AREAS, NR. SUNDBY BRYGGE
3.大湖山庄
4.澳大利亚悉尼-悉尼5号湿地
5，6.芭堤雅希尔顿酒店

1-4.芭堤雅希尔顿酒店
5，6.白香果
7.半岛湖边小屋

Viewing pool ○ 观赏池

1,2.半岛湖边小屋
3.傍花村温泉度假村
4.北京-低碳住家——北京褐石公寓改造设计
5.半岛湖边小屋

1.北京-花石间
2.北京-龙湾别墅和院
3.北京-京都高尔夫中式别墅
4.北京-润泽庄园
5.北京市海淀区-新诗意山居
6，7.北京-泰禾运河岸上的院子
8.碧湖别墅

Viewing pool ○ 观赏池

Viewing pool ○ 观赏池

1，2.碧湖别墅
3.财富公馆
4，5.波士顿－帕梅特谷
6.成都－翡翠城汇锦云天3

1-3.成都美景金山
4.成都双流-和贵馨城
5.成都-玉都别墅庭园

Viewing pool 观赏池

1-3.丹麦哥本哈根-云
4，5.德国柏林-晋城市儿童公园
6.东莞-峰景高尔夫

Viewing pool ○ 观赏池

1，2.东莞-峰景高尔夫
3.东莞市-湖景壹号庄园私家庭
4，5.都柏林－倾斜空间－gal 17
6.芬洛市-马斯河大街—芬洛马斯河岸新城区

Viewing pool ○ 观赏池

41

Viewing pool ○ 观赏池

1. 广州-保利国际广场
2. 广东省-美的总部大楼
3. 复地朗香80-1
4. 广州-保利国际广场
5，6. 广州-凤凰城8号

1. 韩国首尔-West Seoul Lake Park
2. 广州-华南碧桂园
3. 广州-汇景新城别墅
4. 广州-南景圆
5，6.杭州-江南水乡碧水铭苑
7.杭州-江南水乡碧水铭苑
8，9.河源城区-河源东江·首府花园
10.河盛城邦

Viewing pool ○ 观赏池

1,2.荷兰Heemstede-Hageveld Estate
3.荷兰Heerhugowaard-South-Park of Luna
4.荷兰Vught-Parklaan Zorgpark Voorburg
5.墨西哥 - 高科技办公园区 - Tecnoparque Jorge Almanza 1
6.墨西哥 - 马里纳尔可住宅136-12
7.墨西哥-城市与环境设计事务所

Viewing pool ○ 观赏池

1，2.华盛顿－水桌庭园
3.汇景新城
4，5.加利福尼亚－加州威尼斯
6.加利福尼亚－路那达海湾住宅
7.加利福尼亚－马里布海岸豪宅

Viewing pool ○ 观赏池

49

Viewing pool ○ 观赏池

1. 加拿大-Big Sky
2. 江南华府
3. 杰克·埃文斯船港—堤维德岬—期工程
4. 金地格林53号5
5，6. 康涅狄格州郊区住宅-05
7. 兰桥圣菲91号
8. 荔湖城
9. 华南碧桂园燕园

Viewing pool ○ 观赏池

1，2.洛杉矶-维也纳式住宅，加州威尼斯
3.美国加利福尼亚-PV庄园
4.绿洲千岛花园
5.莫斯科 – Green River
6.美国弗吉尼亚州阿什本市-Aquiary户外展示区

Viewing pool ○观赏池

1.秦皇岛-海滩修复工程-015-13 winter view of isle-doted lake-s
2，3.日本茨城县-福冈堰樱花公园
4.加利福尼亚-西亚当斯住宅
5.挪威奥斯陆-Pilestredet公园
6.挪威奥斯陆-Tjuvholmen
7.七宝
8.慕尼黑-施瓦宾花园城市
9.墨西哥墨西哥城-Radial Garden Mexico Historical Center

1,2.日式枯山水庭院－水景植物动物
3-5.瑞士－Enea景观设计公司总部
6.上海－底楼花园

Viewing pool ○ 观赏池

5

6

1. 上海－底楼花园
2. 上海－兰乔圣菲
3. 上海－丽茵别墅
4. 上海－绿城
5. 上海－绿洲江南
6. 上海－绿洲江南园

1. 上海浦东-中邦城市雕塑花园
2，3. 上海市-比利华1
4. 上海市-比利华2
5. 上海市-春天花园
6. 上海市-大华锦绣
7. 上海市-豪嘉府邸别墅花园
8. 上海市-东町庭院

Viewing pool ○ 观赏池

4

5

6

7

8

Viewing pool ○观赏池

1.上海市-豪嘉府邸别墅花园
2,3.上海市-花语墅
4.上海市-锦麟天地雅苑景观设计
5.上海市-圣塔路斯别墅花园
6.上海市-佘山高尔夫夜景

Viewing pool ○观赏池

1-5.上海市-太阳湖大花园设计造园
6.上海市-汤臣高尔夫别墅

Viewing pool ○ 观赏池

1-3.上海市-月湖山庄假山水景
4-6.上海市-新律花园

1. 河源城区-河源东江·首府花园
2, 3. 上海-汤臣高尔夫
4, 5. 上海-万科朗润园
6. 上海-西郊大公馆
7, 8. 上海-香溪澜院

Viewing pool ○ 观赏池

Viewing pool ○ 观赏池

1.上海－香溪澜院
2，3.上海－小别墅
4.上海－新律花园
5.佘山－3号园
6.佘山的梦想

1,2.佘山-东紫园
3.佘山的梦想
4,5.佘山-东紫园
6-8.佘山-高尔夫别墅花园

Viewing pool ○ 观赏池

1-5.佘山－高尔夫别墅花园
6,7.深业.紫麟山
8.深圳—SHENZHEN BAY COASTLINE PARK

Viewing pool ○ 观赏池

1.深圳－万科金域华府－水景轴线
2.沈阳市－汇程庭院
3.深圳－中海大山地
4-7.深圳－卧式摩天楼

Viewing pool ○ 观赏池

3

4

5

6

7

Viewing pool ○ 观赏池

1-3.沈阳-万科金域蓝湾7
4,5.圣地亚哥-拉卡斯塔格伦社区
6.沈阳-万科新里程2

Viewing pool ○ 观赏池

1.顺德－碧桂园
2.台湾－茂顺生活会馆
3.斯塔斯福特
4.太湖高尔夫－泰式水景
5，6.泰国曼谷-Prive by Sansiri

Viewing pool ○ 观赏池

1.唐山 – 唐人起居 – p07
2-6.泰国曼谷–Prive by Sansiri

Viewing pool ○ 观赏池

1.唐山－唐人起居－w01
2.特拉福德滨河长廊
3,4.提香草堂
5,6.天鹅湖
7,8.同润加州

Viewing pool ○ 观赏池

1. 万科-兰乔圣菲
2. 同润加州
3，4. 万成华府
5. 香碧歌庄园
6. 美国俄勒冈-Jon Storm Park
7. 万科朗润园
8. 七宝
9. 新南威尔士帕丁顿-PADDINGTON RESERVOIR GARDENS

Viewing pool ○ 观赏池

1，2.万科朗润园
3.万科棠樾
4.威尔克斯巴里堤防加高河流公地
5.威尼斯
6，7.威尼斯水城别墅花园nEO

1. 五栋大楼屋顶花园
2. 五溪御龙湾
3. 武汉－天下别墅
4，5.西安－世界园艺博览会－greenhouse 3
6，7.西郊一品

Viewing pool ○ 观赏池

91

Viewing pool ○ 观赏池

1.上海-绿城
2-4.西郊一品
5.悉尼-Pirrama公园
6.香碧歌庄园

1. 香格里拉
2. 雪梨澳乡别墅庭院
3. 亚利桑那州-梅萨艺术中心207-12
4. 阳光九九造园
5，6.叶片小斋eaf house
7. 亚洲大酒店

Viewing pool ○ 观赏池

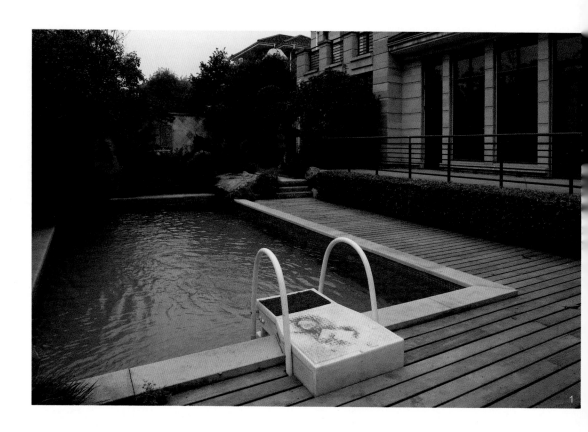

1-3.银都名墅
4.以色列哈尔阿达尔-Har Adar #2住宅
5.云间绿大地
6,7.英国伦敦-现代茶园洋房

Viewing pool ○ 观赏池

1.中国杭州-山湖印别墅
2,3.月湖山庄假山水景
4.芭堤雅希尔顿酒店
5-8.棕榈源园

Viewing pool ○ 观赏池

Viewing pool ○观赏池

1.宗教墓葬文化展示园
2，3.尊
4.半岛湖边小屋
5，6.Brumadinho, Brazil Date of Completion-Burle Marx教育中心
7.爱尔兰-"自我"圣坛

1.上海-绿城
2.上海市-春天花园
3.芭堤雅希尔顿酒店

GARDEN LANDSCAPE DETAIL DESIGN · FOUNTAIN

喷泉+旱喷

　　庭院的意义在于居住者对它所提供的生活方式的认同，在于它所带来的院落"能指"（居住环境）和"所指"（居住者的生活方式、社会交往方式）的和谐关系，而此代码在现代集合住宅中的被扭曲和遗忘引发了居民对院落中亲密的邻里关系的怀念。

　　庭院作为人为的自然空间，它将自然和人工相结合起来，让人产生一种回归自然的向往。庭院中往往常采用小巧、精致的手法，设置植物、院路、亭廊、假山、雕塑、水池等，既表现出个性化的特点，同时给人们带来自然美、人工美的艺术享受。

Agnatus eiumendae voluptamusa custet et demporeribus minverchil mi, unte magnit eum et que doloreh enihili quaspe vendi ium nume sum vita doloria nosam as utendi te solori beribus pa consequi idias maiorem porestiis exerendias as es ut volupit, el earum labo. Itatemp orerenes ipsunt, simusap idustenet explacea et.
con rat doluptatus arum et que plit occum et iunt quam, voluptia parcius dollect aturehentum abo. Neque voluptas seque sam, qui opta volorib ernate lat latem eumet la quis qui ut arit, am reperates dolore.

1,2.Chicago, Illinois-The Park at Lakeshore East
3,4.City Park, Bradford
5,6.挪威奥斯陆-Pilestredet公园

Fountain ○ 喷泉

Fountain ○ 喷泉

1. City Park, Bradford
2. Greenwich Residence, Greenwich, Connecticut
3. Holstebro, Denmark—If it is outstanding, celebrate it twice
4. Los Angeles, CA—Wilmington Waterfront Park
5. Quartz Mountain Residence, Paradise Valley, Arizona
6, 7. South Africa—University of Johannesburg Arts Centre

Fountain ○ 喷泉

1.Xochimilco, Mexico City.-XOCHIMILCO ECOLOGICAL PARK
2.北京-花石间
3.Woody Creek Garden, Pitkin County, Colorado
4，5.北京市-东方普罗旺斯

Fountain ○ 喷泉

1，2.北京市-东方普罗旺斯
3.北京-泰禾运河岸上的院子
4.北京-御墅临枫
5.美国纽约-亚瑟·罗斯平台
6.成都美景金山
7.碧湖别墅
8.成都双流-和贵馨城

Fountain ○ 喷泉

1，2.德国巴伐利亚02007 Waldkirchen花园展
3，4.复地朗香80-1
5.广东东莞-东莞长城世家
6，7.广州-华南碧桂园

Fountain ○ 喷泉

1.广州-保利国际广场
2.韩国首尔-West Seoul Lake Park
3.荷兰阿姆斯特丹-van Beuningenplein
4.加拿大-Big Sky

1.加拿大多伦多-Sugar Beach
2.江西-庭院
3.江西省-爱丁堡二期
4.兰阿姆斯特丹-public square Columbusplein
5.金碧湖畔6号
6.Holstebro, Denmark-If it is outstanding, celebrate it twice

Fountain ○ 喷泉

Fountain ○ 喷泉

1-4.City Park, Bradford
5.曼萨纳雷斯线性公园arg2 © anamuller_300
6.墨西哥-城市与环境设计事务所
7，8.墨西哥 - 高科技办公园区
9.挪威奥斯陆-Pilestredet公园

Fountain ○ 喷泉

1.挪威奥斯陆-Pilestredet公园
2-4.瑞士日内瓦-Renewal of Place des Nations
5，6.挪威-南森公园
7.萨缪尔·德·尚普兰滨水长廊

Fountain ○ 喷泉

1.上海-绿城
2.桑德贝滨水景观,亚瑟王子码头
3,4.上海市-比利华1
5.上海市-春天花园

1.上海-宝绿园
2.上海市-东町庭院
3.上海市-比利华2
4.上海市-春天花园
5.上海市-平江路底楼花园
6.上海市-锦麟天地雅苑景观设计
7.上海市-太阳湖大花园设计造园
8.上海市-佘山高尔夫夜景

Fountain ○ 喷泉

1. 上海市-太阳湖大花园设计造园
2. 上海-汤臣高尔夫
3. 深业.紫鳞山
4. 沈阳市-汇程庭院
5. 圣安东尼奥德克萨斯-Main Plaza Shade Structures
6. 台湾-茂顺生活会馆
7. 泰国曼谷-Prive by Sansiri

Fountain ○ 喷泉

Fountain ○ 喷泉

1，2.同润加州
3.万成华府
4，5.亚洲大酒店
6.武汉－天下别墅
7.西班牙圣塞瓦斯蒂安-AITZ TOKI 别墅

Fountain ○ 喷泉

1.瑞士日内瓦-Renewal of Place des Nations
2.加拿大多伦多-Sugar Beach
3.兰阿姆斯特丹-public square Columbusplein
4.曼萨纳雷斯线性公园arg1 ⓒ anamuller_300
5.美国纽约-亚瑟·罗斯平台
6.瑞士日内瓦-Renewal of Place des Nations

1. 意大利洛迪-漫步花园
2. 悉尼-Pirrama公园
3. 新加坡-City Square Urban Park

GARDEN LANDSCAPE DETAIL DESIGN · WATERFALL

瀑布

庭院的意义在于居住者对它所提供的生活方式的认同，在于它所带来的院落"能指"（居住环境）和"所指"（居住者的生活方式、社会交往方式）的和谐关系，而此代码在现代集合住宅中的被扭曲和遗忘引发了居民对院落中亲密的邻里关系的怀念。

庭院作为人为的自然空间，它将自然和人工相结合起来，让人产生一种回归自然的向往。庭院中往往常采用小巧、精致的手法，设置植物、院路、亭廊、假山、雕塑、水池等，既表现出个性化的特点，同时给人们带来自然美、人工美的艺术享受。

gnatus eiumendae voluptamusa custet et demporeribus minverchil mi, unte magnit eum et que doloreh enihili uaspe vendi ium nume sum vita doloria nosam as utendi te solori beribus pa consequi idias maiorem porestiis xerendias as es ut volupit, el earum labo. Itatemp orerenes ipsunt, simusap idustenet explacea et. on rat doluptatus arum et que plit occum et iunt quam, voluptia parcius dollect aturehentum abo. Neque voluptas eque sam, qui opta volorib ernate lat latem eumet la quis qui ut arit, am reperates dolore.

Waterfall ○ 瀑布

1. Austin, USA Date of Completion-串联式溪边住宅
2-5. Expo 2008 Main Building
6. Salvokop, Pretoria, Gauteng-The Freedom Park

Waterfall ○ 瀑布

1.佛山－天安鸿基花园－天安j
2.广东佛山-山水庄园高档别墅花园
3-5.桂林南溪宾馆
6,7.韩国汉城-ChonGae Canal Restoration Project
8.金地格林53号4

1，2.美国弗吉尼亚州阿什本市-Aquiary户外展示区
3.兰桥圣菲91号
4.沁风雅泾28号
5.挪威-Gudbrandsjuvet -Viewing platforms &&
bridges
6.上海-绿洲江南
7.上海市-万科蓝山

Waterfall ○ 瀑布

Waterfall ○ 瀑布

1. 桑德贝滨水景观，亚瑟王子码头
2，3.上海市-万科蓝山
4.上海市-锦麟天地雅苑景观设计
5.上海-新律花园
6.上海市-月湖山庄假山水景
7.上海-汤臣

Waterfall ○ 瀑布

1,2.深圳-中海大山地
3.圣地亚哥-拉卡斯塔格伦社区
4.深圳-万科金域华府
5.沈阳-万科金域蓝湾
6.Salvokop, Pretoria, Gauteng-The Freedom Park
7.武汉-天下别墅

Waterfall ○ 瀑布

1. 上海市-月湖山庄假山水景
2. 西郊花园
3，4. 西郊一品
5. 威尼斯水城别墅花园
6. 香格里拉09
7，8. 亚利桑那州－石英山居所，534-06

1. 月湖山庄假山水景
2. 张家港 怡佳苑

GARDEN LANDSCAPE DETAIL DESIGN · STREAM
溪流

　　庭院的意义在于居住者对它所提供的生活方式的认同，在于它所带来的院落"能指"（居住环境）和"所指"（居住者的生活方式、社会交往方式）的和谐关系，而此代码在现代集合住宅中的被扭曲和遗忘引发了居民对院落中亲密的邻里关系的怀念。

　　庭院作为人为的自然空间，它将自然和人工相结合起来，让人产生一种回归自然的向往。庭院中往往常采用小巧、精致的手法，设置植物、院路、亭廊、假山、雕塑、水池等，既表现出个性化的特点，同时给人们带来自然美、人工美的艺术享受。

Agnatus eiumendae voluptamusa custet et demporeribus minverchil mi, unte magnit eum et que doloreh enihili quaspe vendi ium nume sum vita doloria nosam as utendi te solori beribus pa consequi idias maiorem porestiis exerendias as es ut volupit, el earum labo. Itatemp orerenes ipsunt, simusap idustenet explacea et. con rat doluptatus arum et que plit occum et iunt quam, voluptia parcius dollect aturehentum abo. Neque voluptas seque sam, qui opta volorib ernate lat latem eumet la quis qui ut arit, am reperates dolore.

1, 2.Austin, USA Date of Completion-串联式溪边住宅
3.Australia-Adelaide Zoo Entrance Precinct
4.Eugene, Oregon, USA-John E. Jaqua Academic Center for Students Athletes
5.Expo 2008 Main Building
6.Lunada Bay Residence, Palos Verdes Peninsula, Southern Californi

Stream ○ 溪流

1. House by the Creek, Dallas, Texas
2, 3. Houston, TX-The Brochstein Pavilion
4. Malinalco House, Malinalco, State of Mexico, Mexico
5. RGA Landscape Architects www.rga-pd.com
6. Barcelona, Spain Date of Completion-Tanatorio Ronda de Dalt花园

Stream ○ 溪流

1.RGA Landscape Architects
2.Tables of Water, Lake Washington, Washington
3.Taiwan, China Date of Completion-日月潭观光局管理处
4-7.阿肯色州小石城-Little Rock Courthouse

1. 澳大利亚-东联高速公路
2. 安徽省-合肥绿地国际花都二期"玫瑰苑"
3，4.半岛湖边小屋
5，6.芭堤雅希尔顿酒店
7.北京-八达岭下的美式小院
8.北京市海淀区-新诗意山居

Stream ○ 溪流

Stream ○ 溪流

1. 北京市通州区-通州北运河河岸景观
2. 北京-泰禾运河岸上的院子
3，4. 碧桂园
5，6. 碧湖别墅
7. 波特兰花园60号
8. 成都美景金山
9. 丹麦，奥尔堡-DANISH CROWN-AREAS, NR. SUNDBY BRYGGE

Stream ○ 溪流

1，2.丹麦，奥尔堡-DANISH CROWN-AREAS, NR. SUNDBY BRYGGE
3-5.德国-柏汉思群岛园
6，7.东莞-峰景高尔夫
8.东莞市-湖景壹号庄园私家庭

Stream ○ 溪流

1,2.埃布罗河畔
3.爱涛漪水园承泽苑16栋
4,5.澳大利亚-东联高速公路
6-8.韩国汉城-ChonGae Canal Restoration Project

Stream ○ 溪流

1-3.华南碧桂园燕园
4.美国弗吉尼亚州阿什本市-Aquiary户外展示区
5.加拿大-东南福溪可持续社区
6.江南华府2
7.江西省-爱丁堡二期

1.上海市-平江路底楼花园
2-4.挪威-南森公园
5.日本茨城县-福冈堰樱花公园
6.挪威-Videseter railings
7-9.挪威奥斯陆-Tjuvholmen

Stream ○ 溪流

1,2.上海-宝绿园
3,4.上海-绿洲江南园
5,6.上海浦东-中邦城市雕塑花园

Stream ○ 溪流

Stream ○ 溪流

1，2.上海浦东-中邦城市雕塑花园
3.上海市-万科蓝山
4，5.上海市-东町庭院
6.上海市-观庭
7.上海市-金地格林

1.上海市-金地格林
2，3.上海市-锦麟天地雅苑景观设计
4.西安 - 世界园艺博览会 - +scape! - flowing gardens
5.深圳-中海大山地
6.圣地亚哥-拉卡斯塔格伦社区

Stream ○ 溪流

Stream ○ 溪流

1.七宝
2.上海市-平江路底楼花园
3,4.西班牙圣塞瓦斯蒂安-AITZ TOKI 别墅
5.香格里拉
6.银都名墅

173

1.阿肯色州小石城-Little Rock Courthouse
2.挪威奥斯陆-Tjuvholmen
3.银都名墅

GARDEN LANDSCAPE DETAIL DESIGN · TERRAIN
地形

　　庭院的意义在于居住者对它所提供的生活方式的认同，在于它所带来的院落"能指"（居住环境）和"所指"（居住者的生活方式、社会交往方式）的和谐关系，而此代码在现代集合住宅中的被扭曲和遗忘引发了居民对院落中亲密的邻里关系的怀念。

　　庭院作为人为的自然空间，它将自然和人工相结合起来，让人产生一种回归自然的向往。庭院中往往常采用小巧、精致的手法，设置植物、院路、亭廊、假山、雕塑、水池等，既表现出个性化的特点，同时给人们带来自然美、人工美的艺术享受。

Agnatus eiumendae voluptamusa custet et demporeribus minverchil mi, unte magnit eum et que doloreh enihili quaspe vendi ium nume sum vita doloria nosam as utendi te solori beribus pa consequi idias maiorem porestiis exerendias as es ut volupit, el earum labo. Itatemp orerenes ipsunt, simusap idustenet explacea et.
con rat doluptatus arum et que plit occum et iunt quam, voluptia parcius dollect aturehentum abo. Neque voluptas seque sam, qui opta volorib ernate lat latem eumet la quis qui ut arit, am reperates dolore.

1. Anchorage, Alaska, USA–Anchorage Museum Expansion
2. Adelaide Zoo Giant Panda Forest
3. Cairnlea, Victoria–Cairnlea
4. Asturias,–运动休闲中心
5. Barcelona, Spain Date of Completion–Tanatorio Ronda de Dalt花园

Terrain ○ 地形

1-3.Australia-Adelaide Zoo Entrance Precinct
4.Austin, USA Date of Completion-串联式溪边住宅

Terrain ○ 地形

1-3.Chicago, Illinois-The Park at Lakeshore East
4,5.Canada-cSimcoe WaveDeck
6.Canberra, Australia Date of Completion-澳大利亚国家美术馆-澳大利亚花园

1.Charleston, USA Date of Completion-查尔斯顿滨水公园
2,3.Expo 2008 Main Building
4,5.Faqra-魔力居所FEJ_9208
6,7.Expo 2008 Main Building

Terrain ○ 地形

1,2.Istanbul, Turkey Date of Completion-荷鲁斯之眼环保建筑
3.London-水域郊野公园游乐区
4.Heerhugowaard-South-月神公园-dephosphatizing pond
5.Houston, TX-ConocoPhillips World Headquarters
6.Jeonnam-小岛上的动物园 ISLAND ZOO - AVIARY INTERIOR 01
7,8.Lakeway Drive - In Redevlopment

Terrain ○ 地形

Terrain ○ 地形

1,2.Linz, Austria-乡村别墅公园和散步广场
3.Lima-绿意来袭
4.Madrid RIO
5.Salvokop, Pretoria, Gauteng-The Freedom Park
6,7.Seattle, Washington-奥林匹克雕塑公园

Terrain ○ 地形

1-3.Monterrey, Mexico Date of Completion-钢铁博物馆
4.Naples-前卷烟厂场重建--MCA_RAM_boulevard entrance
5.Prahova, Romania Date of Completion-Atra Doftana酒店
6.Plenty Road, Bundoora, Victoria-University Hill Town Centre && Wetland

1-3.Taiwan, China Date of Completion-日月潭观光局管理处
4-7.Xochimilco, Mexico City.-XOCHIMILCO ECOLOGICAL PARK

Terrain ○ 地形

Terrain ○ 地形

1，2.阿肯色州小石城-Little Rock Courthouse
3，4.阿姆斯特丹-辛克尔小岛
5-7.埃布罗河畔
8.爱尔兰都柏林-登陆

1.爱尔兰都柏林-观海
2.安徽省-合肥绿地国际花都二期"玫瑰苑"
3.爱尔兰都柏林-登陆
4.安徽省-合肥绿地国际花都二期"玫瑰苑"
5，6.安徽省-合肥政务文化主题公园

Terrain ○ 地形

1,2.盎格滨海路
3.安斯伯利镇区
4.澳大利亚-不莱梅洲际中学
5,6.澳大利亚-库吉港
7.澳大利亚悉尼-Ballast Point 公园

1-3.澳大利亚悉尼-Bondi 至Bronte 滨海走廊
4.白香果
5.百家湖
6.半岛湖边小屋
7.荷兰Velsen-public park Duinpark
8.鲍恩海滨
9.北京-过山车

Terrain ○ 地形

Terrain ○ 地形

1.北京市-竹溪园
2.北京市海淀区-新诗意山居
3.北卡罗来纳州阿什维尔-xueNorth 卡罗莱纳州植物园
4，5.北悉尼地区-BP公司遗址公园

Terrain ○ 地形

1-3.碧湖别墅
4-6.滨海湾花园群
7.兵库县-穹顶多功能网球馆

Terrain ○ 地形

1. 德国巴伐利亚州-德累斯顿动物园 – 长颈鹿园
2. 波士顿 – 帕梅特谷
3. 德国 柏林-南"制革厂"岛
4. 丹麦-哥本哈根的西北公园
5,6. 德国巴伐利亚02007 Waldkirchen花园展
7. 德国巴伐利亚州-伯格豪森的巴伐利亚州园林展 - 城市公园

1.德国-柏汉思群岛园
2-4.德国柏林-莫阿比特监狱历史公园
5.德国慕尼黑-巴伐利亚国家博物馆

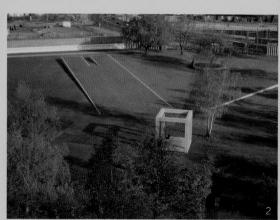

Terrain ○ 地形

1-3.东山墅
4.俄勒冈州-斯蒂恩斯山野马大农场
5.俄勒冈州俄勒冈市-乔恩斯托姆公园
6.都柏林－诺曼底－NORMANDIE. Design by Hugh Ryan. Photo by H Ryan
7.广东东莞-东莞长城世家

Terrain ○ 地形

Terrain ○ 地形

1. 澳大利亚悉尼-Bondi 至 Bronte 滨海走廊
2. 广东省-美的总部大楼
3. 广州-保利国际广场
4. 桂林南溪宾馆
5，6. 广州-凤凰城8号
7. 河源城区-河源东江·首府花园

1.深圳-卧式摩天楼SHA Vanke
2.复地朗香75-4
3.广东省-美的总部大楼
4.荷兰Heerhugowaard-South-Park of Luna
5.荷兰Cuijk-Maaskade Cuijk
6.荷兰Velsen-public park Duinpark

Terrain ○ 地形

5

6

1. 荷兰Vught-Parklaan Zorgpark Voorburg
2. 滨海湾花园群
3，4.荷兰阿姆斯特丹-"Meer-park" Amsterdam
5，6.荷兰海牙-Palace Gardens
7. 湖景壹号别墅庄
8. 华盛顿-互惠中心屋顶花园

Terrain ○ 地形

215

Terrain ○ 地形

1.Anchorage, Alaska, USA-Anchorage Museum Expansion
2.霍恩博格海滨公园
3，4.加拿大，私人公寓，兰乔圣菲
5.加拿大-Big Sky

Terrain ○ 地形

1，2.加拿大魁北克-Canadian Museum of Civilization Plaza
3.挪威阿克斯胡斯-Kjenn市政公园，Lørenskog
4.加拿大魁北克-Core Sample
5.加拿大蒙特利尔-Square Dorchester—Place du Canada

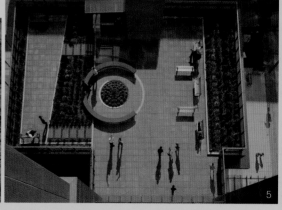

Terrain ○ 地形

1. 马萨诸塞州-波士顿儿童博物馆广场
2，3.美国波特兰-街区
4.美国弗吉尼亚州阿什本市-Aquiary户外展示区
5.美国-哥伦比亚广场
6，7.美国旧金山-Cow Hollow学校操场
8-10.美国马萨诸塞州波士顿-Levinson Plaza, Mission Park

1. 美国明尼苏达州明尼阿波利斯-金牌公园
2. 墨尔本-Elwood海滩景观设计
3，4.美国纽约-亚瑟·罗斯平台
5.广东省-美的总部大楼

Terrain ○ 地形

Terrain ○ 地形

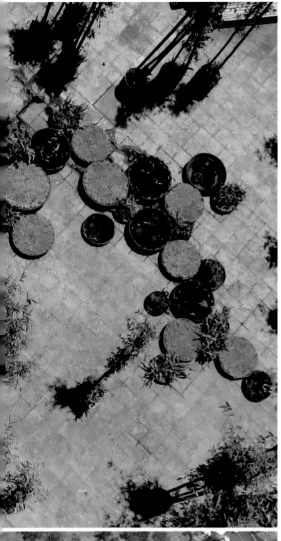

1.墨西哥墨西哥城-Radial Garden Mexico Historical Center
2.墨尔本 - Frankston - frankston aerial
3-5.墨尔本-皇家湿地公园

Terrain ○ 地形

1.墨尔本-儿童艺术游乐园
2-4.慕尼黑-施瓦宾花园城市
5,6.南海市民广场和千灯湖公园
7,8.纽约-龙门广场州立公园

Terrain ○ 地形

1-4.纽约-现代艺术博物馆屋顶花园
5.埃布罗河畔
6-8.挪威-南森公园

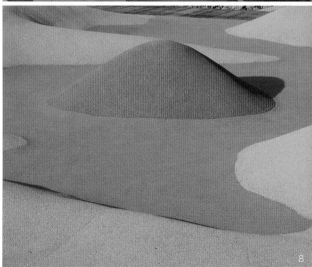

1.盘龙城
2.秦皇岛－海滩修复工程－015-15 board walk-s
3.瑞士日内瓦-Renewal of Place des Nations
4.萨缪尔·德·尚普兰滨水长廊
5.日本茨城县-福冈堰樱花公园

1-3.桑德贝滨水景观，亚瑟王子码头
4.上海-绿城
5，6.上海市-比利华1
7.上海市-比利华2
8，9.上海市-大华锦绣

Terrain ○ 地形

Terrain ○ 地形

1，2.上海市-锦麟天地雅苑景观设计
3-5.上海市-平江路底楼花园
6.上海市-佘山高尔夫夜景

1，2.上海市-新律花园
3.圣地亚哥-拉卡斯塔格伦社区
4.上海市-豪嘉府邸别墅花园
5.上海市-御翠园

Terrain ○ 地形

Terrain ○ 地形

1.深圳-SHENZHEN BAY COASTLINE PARK
2.深圳-万科金域华府-屈曲回折的园中小径
3.深圳-万科金域华府-万科金域华府和院美墅
4.深圳-卧式摩天楼
5.深圳-中海大山地

1. 圣安东尼奥德克萨斯-Main Plaza Shade Structures
2. 圣淘沙跨海步行道
3-5. 台湾-基隆海洋广场

Terrain ○ 地形

1，2.泰国曼谷-Prive by Sansiri
3.上海 – 银都名墅别墅花园
4-6.唐山 – 唐人起居
7.特拉福德滨河长廊

Terrain ○ 地形

1.土耳其-雅勒卡瓦克海滨码头
2,3.西安-"大挖掘"园-地形
4.威尔克斯巴里堤防加高河流公地
5.北京-天竺新新家园
6.土耳其-雅勒卡瓦克海滨码头

Terrain ○ 地形

1. 西雅图 - 山顶住宅
2. 悉尼-奥林匹克公园Jacaranda广场
3. 西班牙伊伦市-Alai Txoko公园
4，5.悉尼-Pirrama公园

Terrain ○ 地形

Terrain ○ 地形

1.新港码头
2-4.新加坡-City Square Urban Park
5.新南威尔士帕丁顿-PADDINGTON RESERVOIR GARDENS
6，7.新西兰-杰利科北部码头漫步长廊，杰利科大道和筒仓公园

Terrain ○ 地形

1-3.新西兰-杰利科北部码头漫步长廊，杰利科大道和筒仓公园
4.休斯敦得克萨斯州-布法罗河口海滨长廊
5.亚德韦谢姆大屠杀博物馆大楼.jpgAerial View
6.以色列贝尔谢巴-BGU University Entrance Square && Art Gallery

Terrain ○ 地形

1，2.以色列-海尔兹利亚公园
3.意大利-Via Regina公共花园
4，5.意大利博洛尼亚-CIRCOLARE, ECOLE DEL RUSCO
6.意大利卢卡-圣·安娜区公园
7.中国杭州-山湖印别墅

Terrain ○ 地形

1.宗教墓葬文化展示园FG
2-4.钟楼海滩
5.英国伦敦-现代茶园洋房
6.钻石提格公园
7.尊

1. 盎格滨海路
2. 慕尼黑-施瓦宾花园城市
3，5. 东山墅
4. 霍恩博格海滨公园

Terrain ○ 地形

1. Austin, USA Date of Completion-串联式溪边住宅
2. 荷兰Heerhugowaard-South-Park of Luna
3，4.墨尔本-儿童艺术游乐园

GARDEN LANDSCAPE DETAIL DESIGN · ARCHITECTURAL SKETCH
建筑小品

　　庭院的意义在于居住者对它所提供的生活方式的认同，在于它所带来的院落"能指"（居住环境）和"所指"（居住者的生活方式、社会交往方式）的和谐关系，而此代码在现代集合住宅中的被扭曲和遗忘引发了居民对院落中亲密的邻里关系的怀念。

　　庭院作为人为的自然空间，它将自然和人工相结合起来，让人产生一种回归自然的向往。庭院中往往常采用小巧、精致的手法，设置植物、院路、亭廊、假山、雕塑、水池等，既表现出个性化的特点，同时给人们带来自然美、人工美的艺术享受。

Agnatus eiumendae voluptamusa custet et demporeribus minverchil mi, unte magnit eum et que doloreh enihili quaspe vendi ium nume sum vita doloria nosam as utendi te solori beribus pa consequi idias maiorem porestiis exerendias as es ut volupit, el earum labo. Itatemp orerenes ipsunt, simusap idustenet explacea et. con rat doluptatus arum et que plit occum et iunt quam, voluptia parcius dollect aturehentum abo. Neque voluptas seque sam, qui opta volorib ernate lat latem eumet la quis qui ut arit, am reperates dolore.

1.圣淘沙跨海步行道
2.波士顿－帕梅特谷
3.土耳其-雅勒卡瓦克海滨码头
4.斯卡伯勒海滩城市设计规划1、4阶段
5.Adelaide Zoo Giant Panda Forest
6.Aley, Lebanon Date of Completion-消失点

Architectural Sketch ○ 建筑小品

Architectural Sketch ○ 建筑小品

1. Austin, USA Date of Completion-串联式溪边住宅
2. Anchorage, Alaska, USA-Anchorage Museum Expansion
3. 8a The Terrace
4, 5. Australia-Adelaide Zoo Entrance Precinct

1-3.Castell D'emporda露天酒店
4.Connecticut Country House, Westport, Connecticut
5.Eugene, Oregon, USA-John E. Jaqua Academic Center for Students Athletes
6.Expo 2008 Main Building

Architectural Sketch ○ 建筑小品

5

6

1.新西兰-杰利科北部码头漫步长廊，杰利科大道和筒仓公园
2.Houston, TX-ConocoPhillips World Headquarters
3.新西兰-杰利科北部码头漫步长廊，杰利科大道和筒仓公园
4.Istanbul, Turkey Date of Completion-荷鲁斯之眼环保建筑
5.以色列哈尔阿达尔-Har Adar #2住宅

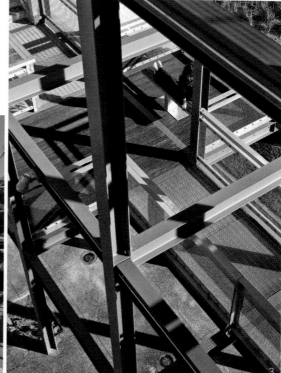

Architectural Sketch ○ 建筑小品

Architectural Sketch ○ 建筑小品

1，2.鲍恩海滨
3.Prahova, Romania Date of Completion-Atra Doftana酒店
4.San Luis Potosí, Mexico Date of Completion-圣路易总体规划
5.Santa Monica-欧几里德公园
6，7.Taiwan, China Date of Completion-日月潭观光局管理处

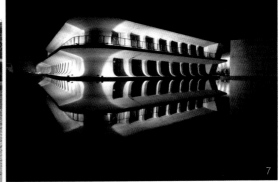

Architectural Sketch ○ 建筑小品

1，2.Urban Play Garden
3.Villa H. St. Gilgen
4.澳大利亚悉尼-Pimelea Play Grounds Western Sydney Parklands
5.澳大利亚悉尼Birchgrove-8a The Terrace庭院

Architectural Sketch ○ 建筑小品

1-3.白香果
4，5.百家湖
6.半岛湖边小屋

1.半岛湖边小屋
2.北京-波特兰
3,4.北京-翠湖——教授的乐园
5.北京-过山车
6.北京-花石间
7.北京-纳帕尔湾

Architectural Sketch ○ 建筑小品

Architectural Sketch ○ 建筑小品

1，2.北京市-东方普罗旺斯
3.北京-新新家园
4.北京市海淀区-新诗意山居
5，6.碧湖别墅
7.兵库县-穹顶多功能网球馆

1.波特兰花园72号
2.成都美景金山
3.成都-玉都别墅庭园
4.城中豪宅
5,6.成都-玉都别墅庭
7.大豪山林21号
8.大湖山庄

Architectural Sketch ○ 建筑小品

1.大湖山庄
2.广东省-美的总部大楼
3.世爵源墅
4.德国慕尼黑-巴伐利亚国家博物馆
5.德国-什未林国家花园展
6，7.东方普罗旺斯（欧式）

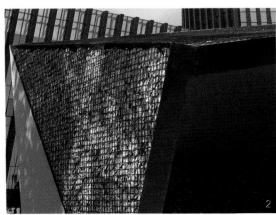

Architectural Sketch ○ 建筑小品

1. 广东佛山-山水庄园高档别墅花园
2. 芙蓉古城
3. 复地朗香80-1
4. 东山墅
5. 广东省佛山市-南海狮山郊野公园景观塔
6. 广州-凤凰城8号

Architectural Sketch ○ 建筑小品

1.广州-金碧花园
2,3.桂林南溪宾馆
4.荷兰Velsen-public park Duinpark
5-7.荷兰阿姆斯特丹-van Beuningenplein

Architectural Sketch ○ 建筑小品

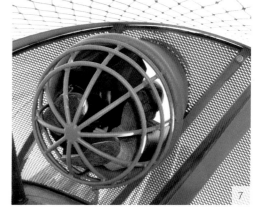

Architectural Sketch ○ 建筑小品

1.上海市-观庭
2.万科棠樾
3.半岛湖边小屋
4.江南华府
5.江西省-爱丁堡二期

1.南海市民广场和千灯湖公园
2,3.乐山市-恒邦·翡翠国际社区
4.新奥尔良-植物园"小屋"
5,6.上海市-佘山高尔夫夜景

Architectural Sketch ○ 建筑小品

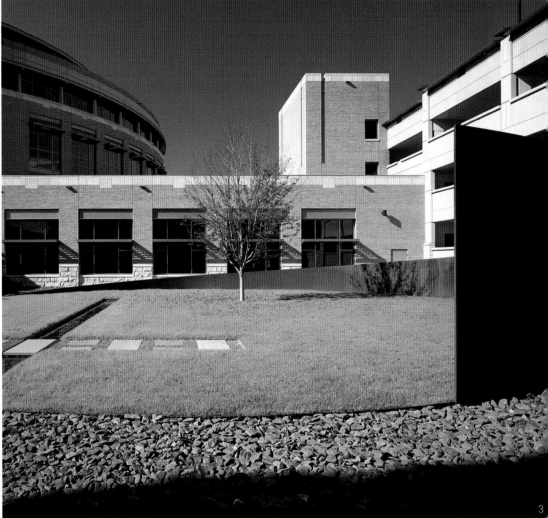

Architectural Sketch ○ 建筑小品

1.美国加利福尼亚-特伦顿大道
2.美国俄勒冈-Jon Storm Park
3.美国明尼苏达州弗里德里-美敦力公司专利花园
4，5.桑德贝滨水景观，亚瑟王子码头

1,2.纽约-龙门广场州立公园
3.纽约-南叉自然历史博物馆
4,5.纽约-探索中心
6,7.挪威-Gudbrandsjuvet -Viewing platforms && bridges

Architectural Sketch ○ 建筑小品

Architectural Sketch ○ 建筑小品

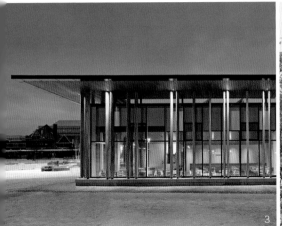

1.挪威-Gudbrandsjuvet -Viewing platforms &&
bridges
2.炮台公园市街区
3.桑德贝滨水景观，亚瑟王子码头
4.日式枯山水庭院 - 入户镇宅树

Architectural Sketch ○ 建筑小品

1.上海浦东-中邦城市雕塑花园
2,3.上海-兰乔圣菲
4.上海-绿城
5.上海市-比华利别墅
6.上海市-比利华2

1，2.上海市-比利华2
3-5.上海市-大华锦绣
6.上海市-东町庭院
7.上海市-观庭

Architectural Sketch ○ 建筑小品

Architectural Sketch ○ 建筑小品

1. 上海市-豪嘉府邸别墅花园
2. 上海市-华庭雅居别墅花园
3，4.上海市-圣塔路斯别墅花园
5. 上海市-太阳湖大花园设计造园

Architectural Sketch ○ 建筑小品

1-3.上海市-太阳湖大花园设计造园
4，5.上海市-汤臣高尔夫别墅
6，7.上海市-御翠园

Architectural Sketch ○ 建筑小品

1.上海－夏州花园
2，3.上海市－月湖山庄假山水景
4，5.佘山－高尔夫别墅
6.深圳－卧式摩天楼

1，2.深圳-卧式摩天楼SHA Vanke 10-03 7232
3.深圳-中海大山地
4，5.沈阳-万科金域蓝湾5

Architectural Sketch ○ 建筑小品

1，2.台湾 - 南投农舍别墅 - sun green
3，4.泰国曼谷-Prive by Sansiri
5，6.泰晤士小镇

Architectural Sketch ○ 建筑小品

Architectural Sketch ○ 建筑小品

1，2.天鹅湖
3，4.天籁园22号
5.万成华府
6.万科-兰乔圣菲

1，2.万科棠樾
3..万科－兰乔圣菲
4.万科棠樾
5，6.威尼斯水城别墅花园
7.五栋大楼屋顶花园

Architectural Sketch ○ 建筑小品

Architectural Sketch ○ 建筑小品

1-3.叶片小斋 leaf house
4-6.以色列哈尔阿达尔-Har Adar #2住宅

Architectural Sketch ○ 建筑小品

1.意大利-Via Regina公共花园
2.云栖蝶谷
3.棕榈湾
4.棕榈源园

1.悉尼-Pirrama公园
2.西雅图－山顶住宅
3.悉尼-奥林匹克公园Jacaranda广场
4.香格里拉
5.小金家

Architectural Sketch ○ 建筑小品

1. 西郊大公馆107号庭院
2. Monterrey, Mexico Date of Completion-钢铁博物馆
3. 新南威尔士帕丁顿-PADDINGTON RESERVOIR GARDENS